Under the Addis Sky

A Coffee Celebration

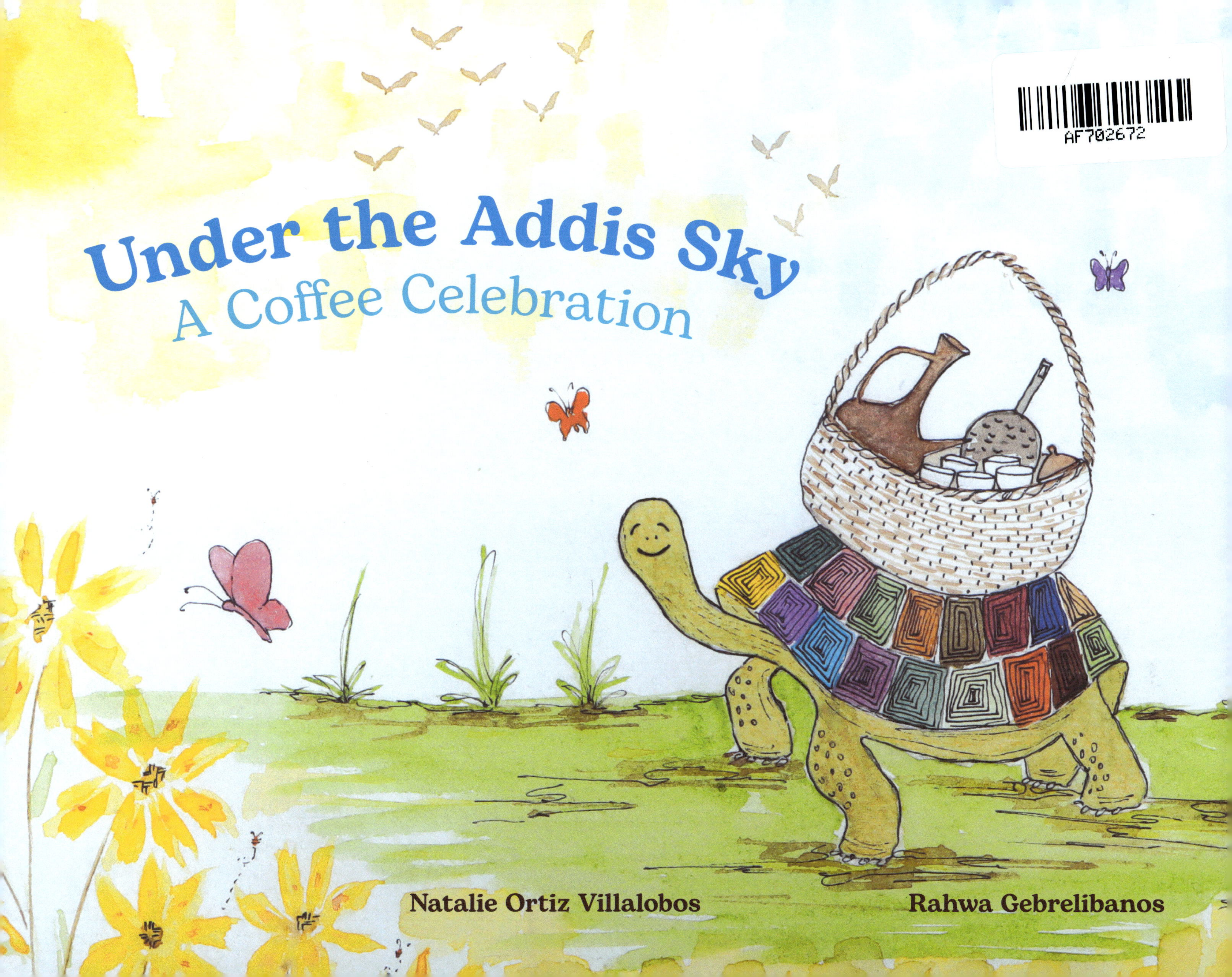

Natalie Ortiz Villalobos

Rahwa Gebrelibanos

For Adina and Farlan, my inspiration

With heartfelt gratitude to everyone who shared their stories, photographs, and memories of the animals living within the British Embassy compound. This book would not have come to life without the wonderful illustrations of Rahwa and the support, encouragement, and generosity of Martha Hardy, David Maynard, Elizabeth Taylor, Yishak Abraham, Jane Kurtz, Birhanu Girma, Bruktawit Abdu, Laura Heaton, Selamawit Mulugeta, Gelila Tesfalidet, Sehin Tewabe, Abinet Teshome, Oliver Russell and many, many friends.

BRITISH COUNCIL

First published in Ethiopia in 2026 by the British Council.
ethiopia.britishcouncil.org

ISBN 978-99990-935-0-7

Under the Addis Sky
A Coffee Celebration

Written by
Natalie Ortiz Villalobos

Illustrated by
Rahwa Gebrelibanos

It was New Year in Addis Ababa. The rainy clouds had drifted away and the sun shone golden again.

High in the Yeka Hills, where trees swayed and wildflowers grew, something lovely was about to happen.

One wise old tortoise was drinking cool water from a dripping spring.

She rolled down a hill and —oops!—

she landed on her back!

Sometimes, the best ideas come when we are upside down.

And just like that, she had a thought...

New Year was here.
And New Year meant...
celebration, a coffee ceremony!

Tortoise called for her most loyal friend, Ibis,
to help her roll back onto her feet.

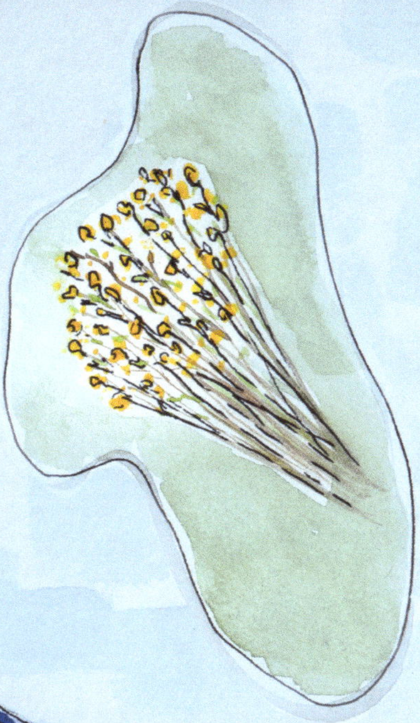

"We shall organise a celebration," said Tortoise, clapping her slow, happy feet.

"Yes!" said Ibis, the tall bird with long legs and a long beak.

"To celebrate the New Year, we must gather everyone. I'll fly across the hills and invite all the animals to a special coffee ceremony!"

Ibis bowed and lifted into the sky.

Flap! Flap! Flap!

Off she went.

First, Ibis visited Monkey, who was sitting high in a tree and licking honey from his fingers. Earlier that morning, a fresh pot of golden honey had been left on a small table near a café, under a lid, very safe.
Or so it seemed.

Pop! Grab! Zip!

Now Monkey hugged the honey pot in the branches, sticky drops shining on his chin.

"Monkey," called Ibis.
"You are invited to the New Year coffee ceremony."

Monkey grinned.
"I'll bring something sweet!"

Next, Ibis flew high past the rocky cliffs, where Eagle was circling above an old stone church.

Thin ribbons of incense smoke drifted into the sky.

Swirl! Rise! Drift!

"Eagle," she called. "Tortoise invites you to the New Year coffee ceremony."

Eagle spread her wide wings.

"I will bring the incense," she said.
"With my wings, I will help the fragrant smoke dance."

Ibis flew low over the hillside. She spotted chickens pecking in the dust.

Peck! Peck! Peck!

Beside them, Wolf snoozed in the shade.

Zzzz! Zzzz! Zzzz!

“Chickens! Wolf!” called Ibis.
“You are invited to the New Year coffee ceremony!”

The chickens fluffed their feathers.
“We will make popcorn!” they clucked.

Wolf flicked his tail.
“I will roast the coffee beans.”

Along a narrow trail between
bushes, a tiny kitten came racing.
She had overheard the others.

“I’ll lead the dancing!” she said.

From nearby came a joyful sound.

Clip-clop! Clip-clop! Clippity-clop!

Horses galloped playfully across the hillside.

"Friends!" called Ibis.
"The coffee ceremony is tonight!"

"We will bring the drums!" they whinnied.

The preparations began.
The chickens popped popcorn.

Pop! Pop! Pop!

The wolf stirred the coffee beans
in a pan.

Crackle! Crackle! Crackle!

Warm, toasty smells floated through the hillside air.

"We make a good team," clucked the chickens.

"Yes," said Wolf softly.
"Some friendships are surprising."

As the sun dipped low, everyone gathered under a big, old tree.

Eagle lit the frankincense.
She fanned her wings gently and the sweet scent drifted through the air.

Monkey placed the honey beside the popcorn on a table made of leaves.

And the wise old tortoise poured
the dark coffee slowly into tiny cups.

Everyone began to dance!

"To friendship, to celebration, to a brand new year."
They all cheered together,

"Melkam Addis Amet! Happy New Year!"

Under the Addis sky, with flowers, popcorn, honey, coffee, and laughter, the animals of the Yeka Hills welcomed the new year.

Wattled Ibis

(Bostrychia carunculata)

The wattled ibis is a pretty, long-legged bird with a curved beak that loves to walk slowly across grass and gardens. You might hear its loud "haa-haa-haa" call echoing in the morning! It pecks for insects and worms, especially after the rains, and sometimes flies in small flocks overhead. This Ibis is only found in Ethiopia and Eritrea.

Tawny Eagle

(Aquila rapax)

The tawny eagle is a big, proud bird that soars high in the sky, scanning the land with sharp eyes. Its feathers are warm brown like the earth, and it loves to glide on sunny winds. Though it is strong and brave, the tawny eagle also likes to sit on tall branches and watch the world below.

Abyssinian Pony

Abyssinian ponies are small but strong horses with shiny coats and lively eyes. They love to run, jump, and play across the fields. These ponies have brave hearts and their legs are perfect for galloping. They are gentle friends who love teaching children how to ride, and their favourite snacks are sweet, crunchy carrots.

Grivet Monkeys

(Chlorocebus aethiops)

Grivet monkeys are clever little creatures who love to jump, climb, and explore. Their long tails help them balance as they leap between branches and move quickly through the trees. They eat fruits, leaves, and seeds, and enjoy playing together.

African Spurred Tortoise

(Centrochelys sulcata)

Tortoises live a very long time, sometimes longer than grandparents! They carry their homes on their backs and stay safe inside their hard shells. They eat leaves, grass, and flowers, and they move slowly. If you see a tortoise on its back, ask an adult for help, as tortoises cannot turn themselves over.

African Golden Wolf

(Canis lupaster)

African golden wolves are watchful quiet animals who roam across plains and hills. Their golden fur helps them blend into the land as they move softly in search of food like small animals, fruits, and insects. They often live in family groups, calling to one another with soft howls at sunset.

Ethiopian Coffee Ceremonies

In Ethiopia, making coffee is an ancient ritual of hospitality and togetherness. First, green coffee beans are roasted over hot coals until they smell warm and nutty. Then, the smoking pan is passed around so everyone can enjoy the wonderful aroma. Next, the coffee is brewed in a beautiful clay pot called a jebena. It is poured into tiny cups three different times.

The first, Abol, is the strongest.
The second, Tona, is weaker.
The third, Baraka, is the weakest and is said to bring a blessing!

While everyone sips together, popcorn, bread, and barley are shared and sweet-smelling incense fills the air. It is a time to slow down, talk, and enjoy being with family and friends.

Ethiopian New Year
(Enkutatash)

Enkutatash happens at the end of the rainy season, when the sun shines bright and the world feels fresh and new. On this special day, people wear and decorate with yellow flowers called Adey Abeba, which bloom at that time of year. Families and friends celebrate with songs, prayers, and food.

Adey Abeba (Bidens macroptera)

Adey Abeba is a bright yellow flower that only grows in the highlands of Ethiopia. It blossoms after the rainy season, covering fields and hills with golden colour. The flower is a joyful sign of the Ethiopian New Year, reminding people of renewal, sunshine, and new beginnings.

Eskista shoulder dance

Eskista is a lively traditional dance from Ethiopia, known for its joyful shoulder movements. Dancers shake and roll their shoulders, chest, and neck to the rhythm of music and drums. The movements can be fast or gentle, telling stories through the body. Eskista is often danced during celebrations, bringing people together in laughter and rhythm.

About the Story

This book was inspired by many walks with my two young children. Together, we paused to observe flowers, birds, tortoises, monkeys, and other animals, sharing in the awe, wonder, and connection that nature can bring. Their curiosity and delight in the world around them planted the first seeds of this story.

The story then grew from the photographs and animal anecdotes shared by families and staff connected to the British Embassy compound in Addis Ababa. Inspired by these contributions, I created a children's story that celebrates the unique wildlife of this place, the rich cultural traditions of Ethiopia, and the joy of sharing a cup of freshly roasted 'buna' with popcorn while frankincense infuses the air.

At its heart, this book is a celebration of community, culture, imagination, and the connections that bind us to one another and the natural world. My hope is that it will be enjoyed by children across Ethiopia, as well as by those living farther away who carry a connection to this beautiful place. May it inspire curiosity, kindness, and a love for the remarkable wildlife and traditions that make Ethiopia so special.

Story Guide - Questions, Vocabulary & Activities to Bring the Story to Life

Book Introduction & Assessing Prior Knowledge:

Ask the children about how they celebrate New Year. Do they have a coffee ceremony? If so, who is there? What types of foods do they eat? What activities do they do?

Vocabulary to Consider / Pre-teaching of Skill:

Tell the children the meaning of the word loyal. Being loyal is an important characteristic of a friend. Ask if they have friends who are loyal. How do they show loyalty?

Picture Walk:

Read the title of the story to the children. Ask them to look at each of the pictures. They can discuss with a partner, or a small group, what they see and what they think might be happening in the story as they look at each page. Ask them to talk about it with the whole group.

Suggested Questions to Ask During Reading:

- How did Ibis solve Tortoise's problem and show she is a loyal friend?
- Does Tortoise have a lot of friends?
- What will Monkey bring?
- Where did he get the honey?
- What will each animal bring to the party?
- Why is the friendship between the chickens and Wolf surprising?

Questions to Ask After Reading:

- The animals were all very different. How were they different?
- Which friendship was surprising?
- Have you ever made friends with someone who was different from you? What did you have in common?
- How did the animals make sure everyone felt welcome?
- Each animal had different strengths and ways of helping. Why is it important that everyone contributes in their own way?
- Who was your favorite character? Why?

Movement Activity:

Ask the kids to mimic the actions of the animals.
Imagine that....

- You are old and wise like a tortoise. Move slowly.
- You have a long beak like an ibis. Stretch your neck forward and back.
- You swing through the trees like a monkey. Swing your arms.
- You fly through the sky like an eagle. Flap your arms.
- You peck at the ground like a chicken. Nod your head.
- You rest like a wolf. Curl up and be still.
- You dance like a kitten. Roll your shoulders.
- You are a horse. Click your tongue and stomp.

ethiopia READS

This guide was developed in collaboration with Ethiopia Reads, an organisation that empowers children through literacy by providing access, training, advocacy, and outreach in communities throughout Ethiopia. **ethiopiareads.org**

Natalie Ortiz Villalobos - Author

Natalie, originally from Costa Rica, has spent much of her life exploring different corners of the world and the stories they hold. She has been living with her family in Ethiopia for several years and has grown to love the country's people, culture, and natural beauty. Natalie is a multidisciplinary designer, yoga teacher, and doula guided by a deep appreciation for nature. Through her storytelling, she hopes to inspire curiosity, wonder, and care for our world.

Rahwa Gebrelibanos - Illustrator

Rahwa is a multidisciplinary artist, sculptor, photographer, and fashion designer based in Addis Ababa. She studied Sculpture at the Alle School of Fine Arts and Design, Addis Ababa University, where she is an assistant lecturer. Through sculpture, Rahwa explores the human figure and the psychology of play. Blending fashion and sculptural practice, she examines how wearable forms influence movement, emotion, perception, and human interaction. Alongside her artistic practice, Rahwa is the mother of three young children, whose creativity and sense of play continue to inspire her work.

www.ingramcontent.com/pod-product-compliance
Lightning Source LLC
Chambersburg PA
CBHW042005180926
51764CB00033B/302
9789999093507